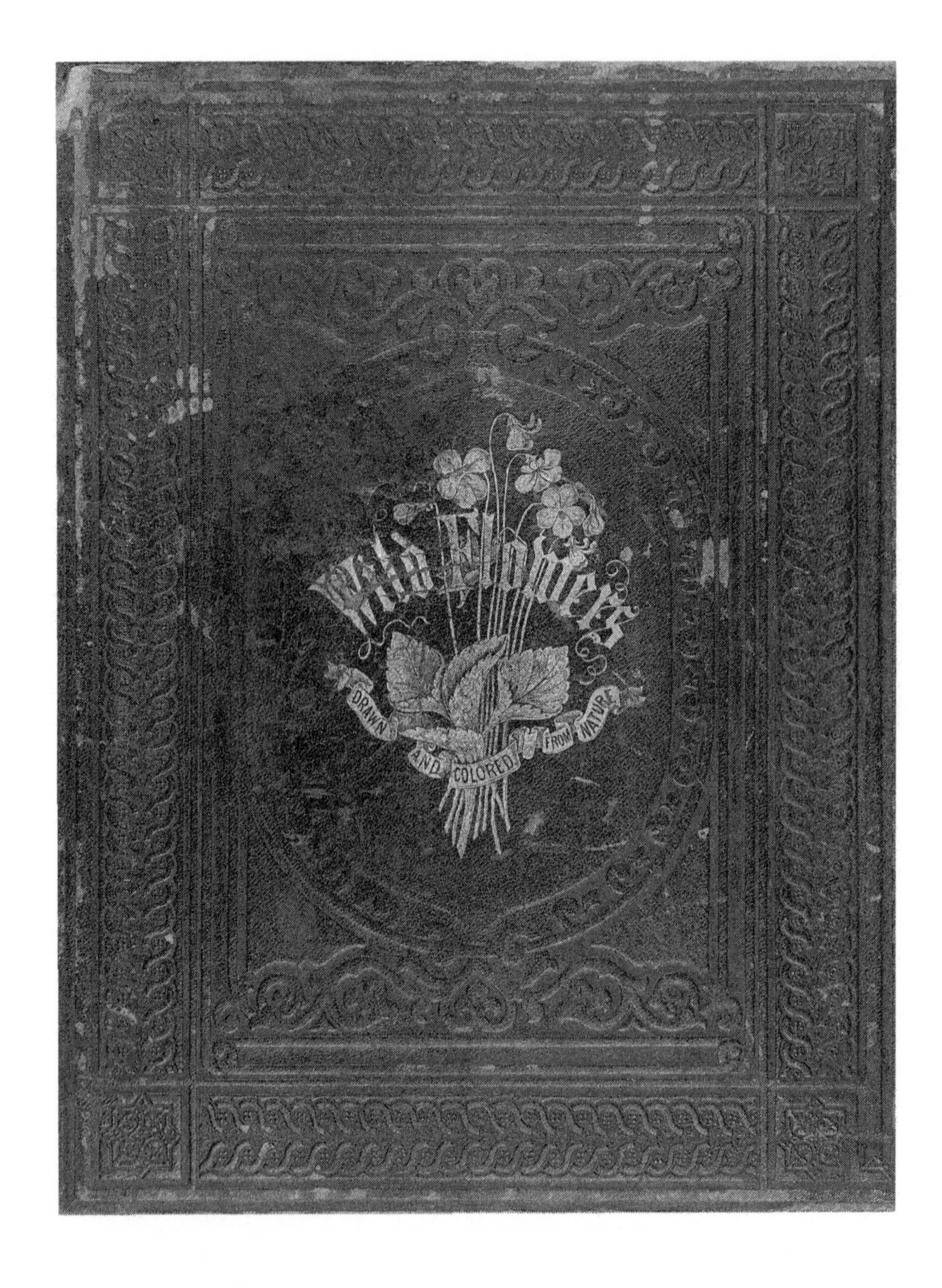
Wild Flowers
DRAWN AND COLORED FROM NATURE

WILD FLOWERS

DRAWN AND COLORED FROM NATURE.

BY

MRS. C. M. BADGER.

WITH AN INTRODUCTION.

BY

MRS. L. H. SIGOURNEY.

WHO DOES NOT LOVE A FLOWER?
ITS HUES ARE TAKEN FROM THE LIGHT
WHICH SUMMER'S SUN FLINGS PURE AND BRIGHT,
IN SCATTERED AND PRISMATIC HUES,
THAT SHINE AND SMILE IN DROPPING DEWS;
ITS FRAGRANCE FROM THE SWEETEST AIR,
ITS FORM FROM ALL THAT'S LIGHT AND FAIR.—
WHO DOES NOT LOVE A FLOWER?

BRAINARD.

NEW YORK:
CHARLES SCRIBNER, 124 GRAND STREET.
LONDON:
SAMPSON LOW, SON & CO., 47 LUDGATE HILL.
1859.

JOHN F. TROW,
PRINTER & STEREOTYPER,
377 Broadway

LIST OF FLOWERS.

CORNUS FLORIDA . . . . DOGWOOD.

A small tree, whose large showy flowers give a beautiful appearance to our woods in May.

Class, Tetrandria: *Order*, Monogynia: *Natural Order*, Cornaceæ.

GERANIUM MACULATUM . . . . WILD GERANIUM.

This light pretty flower, with its downy leaves, and slender stalk, we find early in May, along the woodside, and by the fences; and often its purple head is seen above the tall grass in June.

Class, Monadelphia: *Order*, Decandria: *Natural Order*, Geraniaceæ.

SILENE VIRGINICA . . . . GROUND PINK.

A modest little companion of the Geranium and Columbine.

Class, Decandria: *Order*, Trigynia: *Natural Order*, Caryophyllaceæ.

EPIGÆA REPENS . . . . TRAILING ARBUTUS.

Named by the Pilgrims of New England, *May-Flower*;—said to have been the earliest found, after their first dreary winter in the New World.

Class, Decandria: *Order*, Monogynia: *Natural Order*, Ericaceæ.

LIST OF FLOWERS.

VIOLA CUCULLATA . . . . . HOODLEAVED VIOLET.

This little favorite is one of the earliest flowers of Spring, growing abundantly in meadows and grassy woods.

Class, Pentandria: *Order*, Monogynia: *Natural Order*, Violaceæ.

AQUILEGIA CANADENSIS . . . . WILD COLUMBINE.

An early May Flower, found on the rocky hill-side, and in the most barren soil,—a brilliant contrast to the pale flowers around it.

Class, Polyandria: *Order*, Pentagynia: *Natural Order*, Ranunculaceæ.

CLAYTONIA VIRGINICA . . . . . SPRING BEAUTY.

A lovely child of May,—modest and unassuming,—half hidden in the grass, along the sides of the deep tangled wood.

Class, Pentandria: *Order*, Monogynia: *Natural Order*, Portulacaceæ.

AZALIA NUDIFLORA . . . . . WILD HONEYSUCKLE.

A shrub, in bloom through May, the flowers varying in color from pale pink to deep crimson. It chooses a dry soil, and is abundant in most of our woods.

Class, Pentandria: *Order*, Monogynia: *Natural Order*, Ericaceæ.

PYRUS MALUS . . . . APPLE BLOSSOM.

Class, Icosandria: *Order*, Polygynia: *Natural Order*, Rosaceæ.

LILIUM CANADENSE . . . . YELLOW LILY.

A showy flower, adorning our fields and meadows in July.

Class, Hexandria: *Order* Monogynia: *Natural Order*, Liliaceæ.

CAMPANULA ROTUNDIFOLIA . . . . . HAREBELL.

Found in shady places, and by rocky streams in August.

Class, Pentandria: *Order*, Monogynia: *Natural Order*, Campanulaceæ.

LIST OF FLOWERS.

ROSA SUAVEOLENS . . . . SWEET-BRIER.

Blossoms in June and July, about our woods, and along the road-sides. The color commonly pale, but often deep pink.

Class, Icosandria: *Order*, Polygynia: *Natural Order*, Rosaceæ.

LIRIODENDRON TULIPIFERA . . . . . TULIP-TREE.

A large and splendid forest tree, blossoming abundantly in June.

Class, Polyandria: *Order*, Polygynia: *Natural Order*, Magnoliaceæ.

KALMIA LATIFOLIA . . . . . MOUNTAIN LAUREL.

One of the most elegant shrubs in our country, flowering early in June; and remains of its beauty are to be found in July.

Class, Decandria: *Order*, Monogynia: *Natural Order*, Ericaceæ.

ROSA PARVIFLORA . . . . WILD ROSE.

Abundant through all the Summer months.

Class, Icosandria: *Order*, Polygynia: *Natural Order*, Rosaceæ.

ERIGERON STRIGOSUM . . . . DAISY.

A troublesome weed to farmers, but a favorite flower with children.

Class, Syngenesia: *Order*, Polygamia Superflua: *Natural Order*, Compositæ.

RANUNCULUS ACRIS . . . . BUTTERCUP.

A bright golden flower, familiar to every child; in bloom through the Summer.

Class, Polyandria: *Order*, Polygynia: *Natural Order*, Ranunculaceæ.

TRIFOLIUM PRATENSE . . . . RED CLOVER.

In flower from June to September.

Class, Diadelphia: *Order*, Decandria: *Natural Order*, Leguminosæ.

LIST OF FLOWERS.

RHODODENDRON MAXIMUM . AMERICAN ROSE BAY.

A large and elegant evergreen shrub, flowering in June and July.

Class, Decandria: *Order*, Monogynia: *Natural Order*, Ericaceæ.

ASCLEPIAS TUBEROSA . . . . . BUTTERFLY-WEED.

A bright orange flower, growing in thick clusters among the grass, in dry fields through August.

Class, Pentandria: *Order*, Digynia: *Natural Order*, Asclepiadaceæ.

LILIUM PHILADELPHICUM . . . . . . WOOD LILY.

Found in woods and fields in July and August.

Class, Hexandria: *Order*, Monogynia: *Natural Order*, Liliaceæ.

ORCHIS PYSCODES . . . . PURPLE FRINGED ORCHIS.

A beautiful and delicate flower, adorning the meadows in July.

Class, Syngenesia: *Order*, Monogynia: *Natural Order*, Orchidaceæ.

LOBELIA CARDINALIS . . . . CARDINAL FLOWER.

Common in wet places, and along the margin of streams, in July and August,—often called Indian Feather.

Class, Pentandria: *Order*, Monogynia: *Natural Order*, Lobeliaceæ.

ACER RUBRUM . . . . RED MAPLE.

A forest tree, covered with bright crimson flowers early in April, and sometimes in March, but is unsurpassed in variegated brilliancy, when its leaves are touched by the early frost.

Class, Polygamia: *Order*, Diœcia: *Natural Order*, Aceraceæ.

GENTIANA CRINITA . . . . . . FRINGED GENTIAN.

The last blooming visitor of the woods, standing among the withered leaves, alone in its beauty.

Class, Pentandria: *Order*, Digynia: *Natural Order*, Gentianaceæ.

POETRY.

INTRODUCTION.

Gems, the richest of the vale,
Violets blue, and daisies pale,—
Blossoms nursed by dews that weep
While their fringed eyelids sleep
Lightly on the rocky steep,
Or in Summer's curtained dells
Where the crystal brooklet swells,—
Autumn's leaflets, red and sere,
Life-blood of the parting year,
Mix'd with thrills of song are here.—
Where in single group combined,
Such attractions can ye find?—
—One fair hand, hath skill'd to bring
Voice of bird, and breath of Spring,
One fair hand, before you laid
Flowerets that can never fade,—
While you listen, soft and clear
Steals her wind-harp o'er your ear,—
While you gaze, her buds grow brighter,—
Take the book, and bless its writer.

L. H. S.

New York, Sept. 16, 1856.

THE TRAILING ARBUTUS, OR MAY-FLOWER.

WHEN hath May's sun or April's softening shower,
Warm'd into life a less assuming flower
Than simple me, in pale pink vesture drest,
While close to earth my lowly head finds rest?
But yet, enough might my experience speak,
To tinge with pride a more expanded cheek.
Ere art or science on this land had smiled,
When these fair fields were but a forest wild,
My little race of blossoms, pure and sweet,
Were press'd beneath the dusky Indian's feet.
And when, with weary hearts, the Pilgrim band
Raised their first prayer from Plymouth's frozen strand,
I lay for Spring's return, their eyes to greet,
Concealed from view by nature's winding-sheet.
I was the flower their waiting eyes first found,
Exhaling my faint perfume from the ground;
And with my little downy stalk I crept
On the first grave o'er which their fond hearts wept.
They hailed, as promise of a brighter hour,
My blushing face, and call'd me their MAY-FLOWER.
As then I smiled, with early May dews wet,
In dark-browed maidens' glossy locks of jet,
I now am wreathed amid the sunny curls,
Or braided locks of sweet New England girls.
I waiting stand, when April's parting ray
Is lost in darkness, and the blushing day
Looks forth on groups of lovely maidens straying
O'er groves and fields, with happy hearts, "a Maying;"
With their gay shouts, the budding forests ring,
And me they hail, the first sweet flower of Spring.

SPRING FLOWERS.

LET those who choose love Winter's glow,
And feel regret to have him go,
And look with rapture on the snow,
So still and lightly falling.
I know it is a pretty sight,
To watch its fall, so pure and white;
But yet, although it looks so bright,
Its coldness is appalling.

The flakes that best my spirit please,
Are showered in May,—by gentle breeze,
Brushing across the apple trees,
When robed in snowy flowers.
This is the lightly falling snow,
And this the kind of wind to blow,
For which I would the joys forego,
Of all the wintry hours.

That May-day scene e'en now I see,
Those pearly flowerets dancing free,
And hear ring out that laugh of glee,
From young and happy voices.
E'en now I wander by the stream,
And in resplendent beauty seem
To view the whole as in a dream,
And all my heart rejoices.

I see the banks with violets spread,
And Columbine's bright nodding head,
That up the steep my footsteps led,
In happy fond devotion.

SPRING FLOWERS.

I see the wild-pinks dot the ground,
And bright spring-beauties all around,
And little wind-flowers that abound,
With their light fluttering motion.

Those happy days have passed away;—
But though it is not always May,
The other months bring flowers as gay,
And skies as blue and beaming;
And other groves whose whispering trees
Will tell us tales as sweet as these,
And little openings by the breeze,
Will let in light as gleaming.

Then let the rich who can command
The gold from California's land,
Buy their exotics, tall and grand,
And set in costly vases.
Give me the little flowers that blow,
Along the banks where streamlets flow,
Or such as on the hill-top grow,
Or in the shadiest places.

To these 'tis given to impart
The joy that makes the tear-drop start,
And these have power to move the heart,
E'en of the rudest peasant;
For while he views the blooming ranks
Of violets all along the banks,
He looks above and gives God thanks,
Who makes his paths so pleasant.

They caught the Saviour's radiant eye,
And mingled with his counsels high,
And since he left them for the sky,
They still live in his story:
Behold these careless flowers! he said,
And let thy heart on God be stayed;
For not like these was e'er arrayed
The king in all his glory.

THE VIOLET.

THIS little flower, so sweet and wild,
Is nature's fairest, simplest child,
With whispers soft, May's earliest breeze,
Seeks for it, under budding trees,

By sunny banks and waters cool,—
The gentlest bud in beauty's school.
On wings of love the evening flew,
To bathe it in her crystal dew.

Gray morning tripped on silver feet,
Its earliest blush and smile to meet,
And noon, bright noon, his clear rays shed
In dazzling lustre round its head.

It is the little maiden's pet,
It is the poet's favorite;—
To name it, is to touch a spring,
Which moves each tender bosom string

To music, such as childhood hears,
And love recalls in after years.
'Tis like the bleating of the lambs,
Or mellower voices of their dams;

'Tis like the tinkling heifer's bell,
Or noon-day horn's clear distant swell,
Or like the cooing of a dove,
Or like the gentle voice of love;

THE VIOLET.

'Tis like the stream o'er pebbles flowing,
Or trees through which the breeze is blowing,
Or like the hum when dear friends meet,
And like to every thing that's sweet.

Nor can the memory e'er depart,
Which holds this flower in many a heart,
And who is there that would forget,
The modest blue Wild Violet.

THE SPRING BEAUTY.

I TROD the green-wood's flickering shade,
Where Spring her earliest gems display'd,
And hoarding them, as on I pass'd,
I deem'd each lovelier than the last.
But when this sweet SPRING BEAUTY rose,
I said, with thee, my search shall close;
For who can view this lovely flower,
Nor wake to beauty's silent power?
Its opening petal, soft and sleek,
As glow of love o'er childhood's cheek,
Its buds, like smiles, that nestle in
The rosy mouth and dimpled chin.
I love it, for the poorest child,
May gather of its sweetness wild.
You find it not in garden bower,
The wild-wood owns the lovely flower.

Wide may the eyes of wonder stare,
To view what skill of man may dare,
Where winds of heaven in fragments strip,
And scatter wide, the noblest ship.
Far down beneath the angry wave,
Where rests the stillness of the grave,
Now do his boldest plans essay,
Between two worlds a path to lay,
Where thoughts from each to each may run,
Like rays of light from yonder sun.
Then may the friend fond words repeat,
And distant love the message greet,

THE SPRING BEAUTY.

Before the swiftest wind can blow,
To lay the staggering topmast low,
Whisper'd through ocean's darken'd cell,
Nor mermaids hear the tales they tell.
And yet, with all his skill and power,
Man could not make this simple flower.

THE PINK AZALIA, OR WILD HONEY-SUCKLE.

Sweet flower, whose rustic beauty glows,
Fair as the blush upon the rose,—
Whose name is sweet, whose home is wild,
The valley's and the mountain's child.
I love it for the choice it made,
To blossom in the quiet shade,
I love it for the generous show
With which its honey'd branches blow;
It yields them not in stinted measure,
To give its fond admirers pleasure.

Along the winding paths that lead
Where cattle browse and lambkins feed,
And in the dry and stony soil,
Which frowns upon the farmer's toil,
And by the moss-crown'd rocky ledge,
And by the forest's sunny edge,—
Where the bright oriole soothes to rest
Her young within her pendant nest,—
And by the cool stream's rippling flow,
The tall Wild Honeysuckles grow.

Then let me leave the crowded street,
Where wealth and want and folly meet,
And go where balmy breath of May
Hath made the fields and forests gay,—

THE PINK AZALIA.

O, let me wander where it grows,
And pluck the first sweet bud that blows.
A single blossom would repay
The gloom of many a winter day.
But O, how pure is my delight,
When thousands of them greet my sight.

TO THE WILD COLUMBINE.

SWEET COLUMBINE, what happy thought
Is by thy lovely image brought;
Thy nodding head says, "Come and see
The place where thou hast play'd with me."
And go with thee I gladly will,
The pebbly path is winding still,
The house is standing by the wood,
Where in my childhood's days it stood.
The poplar, firm, erect, and tall,
Beside the rough and ragged wall,
Though the most soldier-like of trees,
Stands shivering from the slightest breeze.
The little mount hath tales to tell,
That always please my fancy well;
While from the hollow just beyond,
Comes music from the shaded pond,
Where harmless frogs, in safe retreat,
Their tale monotonous repeat.
And in the quiet Sabbath air,
I hear the bell's loud call to prayer,
As by long travel half-subdued,
Its tones thrill through the leafy wood.
I hear the cattle's distant low
Come swelling from the vale below.
The geese still gabble,—as they eat
With slanting bill the herbage sweet,—
The story which they all commend,
But none save geese can comprehend.

And still the sea, its hollow roar
Is sending from the far-off shore.
Those little flowers nod on the hill,
And children climb to pick them still,
And there the same broad oak will spread
Its giant arms and cooling shade.
Nor fear I that old fence will fail,
With stony base and top of rail;
I'll mount it, and a single bound
Will land me on the flowery ground.
Anemone, and Buttercup,
Attract me not, for farther up,
Among the stones and creeping pines,
I see the scarlet COLUMBINES.

My life has been a changing dream
Of what things are, and what they seem;
But never shone there brighter hours,
Than when I play'd with those sweet flowers.
And all its changes soon will pass,
Like moving shadows o'er the grass,
Yet might its sun shine, bright as then,
I would not live it o'er again,
For brighter blossoms in the sky,
Bloom fresh and fair, and never die.

SUMMER FLOWERS.

Now wreathe your garlands, bind your posies,
This is the favored time for roses;
Leave city walls and palace gates,
And go where lovely Nature waits
To welcome you with open arms,
And proffer all her blooming charms.
She leads you where the Sweet-brier grows,—
To hedges, bright with Mountain Rose;
She beckons from the shady grove,
Where wild birds sing their songs of love,
And charm you with their chattering noise,
Responding to each other's joys.
She woos you with her softest breeze,
And waves a welcome from the trees;
She leads you to the glassy lake,—
Through Laurel shrubs, and tufted brake,
Where queenly Rhododendron seeks
To view her fair reflected cheeks,—
Whose rippling waters take delight
In throwing back her blushes bright,
While smiling on the silvery sheen,
From out her bowers of evergreen.
We'll follow still, where Nature leads,
Through shady woods, and flowery meads,
To groves, where on the tree-top high,
Fair Tulip flowers attract the eye,
Which deign to bend a lowly glance
Upon the bank where Blue-bells dance.

SUMMER FLOWERS.

And where green meadows broad denote,
That Orchis' silken fringes float.
Then, to the spreading vale we'll go,
Where Buttercups and Daisies grow,
View the Asclepias e'er they writhe
Beneath the mower's sharpened scythe.
And, like the little busy rover,
We'll revel in whole fields of clover,
While justly we divide the shares,
The fragrance ours, the nectar theirs:
Then wend our way by streamlet's edge,
Where mid the brambles, weeds and sedge,
In sweet companionship together,
Dwell Arrow-head and Indian Feather.
Where fields of waving grass are spread,
The LILY bows her graceful head,
We'll venture through the tall green mowing,
And pluck the beauty while 'tis growing,
And keep it, to refresh our sight
Upon some stormy winter night.

WILLIE'S FLOWERS.

'Twas in the month that follows May,
That little Willy ran away,
And his sweet name was echoed shrill,
From every grove and plain and hill.

His mother took her darling boys,
Far from the city's dust and noise,
That they in shady groves might play,
And pick wild flowers and smell the hay.

One morning when the sun was high,
And not a cloud obscured the sky,
When every thing in nature smiled,
She missed one little roguish child.

A narrow bridge beneath the hill,
A clear stream spanned, that turned a mill,
And little Willy loved to look
Into this running laughing brook.

And now the mother, almost wild,
Ran calling on her straying child,
As flowers dropped from his little hand,
His tiny foot-prints in the sand,

Lead where the sun in dazzling beams,
Upon the treacherous water gleams.
But joy! she sees his sunny head,
Beneath a branching oak's dark shade.

And when she gave a joyful shout,
The little fellow turned about,
His apron white, was running over,
With DAISIES, BUTTERCUPS and CLOVER.

The darling feared he was to blame,
And some excuse he needs must frame,
So, with a lovely little smile,
Eyeing his pretty flowers the while,

"Me run, till Willy almost faint,
To pick sweet flowers for Ma-ma paint."
What mother but would press with joy,
Close to her heart the wandering boy?

But when he learned how sad she'd been,
He promised ne'er to stray again.
That honest face, the mother knew,
Told nothing that he would not do.

The rest of that day's happy hours,
She spent in painting Willie's flowers,
And never since, e'en to this day,
Has little Willy run away.

THE SWEET-BRIER.

I'M but a simple way-side brier,
 Yet, by the dusty ways,
The weary traveller's grateful smile
 Gives me my meed of praise.

The shouting school-boy, bounding on,
 Lured by my fragrance, lingers,
Nor frowns, when, gathering of my sweets,
 I prick his venturous fingers.

I smile on many a merry group,
 Along the road that passes,
And rival in their morning bloom,
 The happy rosy lasses.

My scarlet fruit, when flowers are gone,
 The little maiden strings,
And o'er her snowy bosom pure,
 The ruby necklace flings.

Where fair hands round the cottage door
 Have trained my glossy leaves,
My long and slender boughs I stretch
 To meet the bending eaves.

And, often in the hallow'd spot,
  Where pale the mourner weeps,
I breathe my fragrance o'er the grave
  Where youthful beauty sleeps.

Thus, year by year, my little charms
  I cheerfully impart,
Pleased, when I check a sorrowing tear,
  Or cheer one lonely heart.

THE TULIP TREE BLOSSOM.

Daily beneath our careless feet,
We tread the verdant sod,
And at each step such treasures meet,
As come from none but God.

They grow along the way-side bank,
Each sunbeam gives them birth,
They gild with light our pilgrim path,—
We call them stars of earth.

Above the earth, far, far away,
Stretches an azure field,
And every rood of all that ground,
Doth countless beauties yield.

When slowly and with noiseless tread,
Descend the dews of even,
They shine to cheer our nightly way,—
We call them flowers of heaven.

Between the blossoms of the sky,
And stars that strew the earth,
Blooms what I call a child of air,—
A flower of lofty birth.

The birds that gaily soar for heaven,
To rest the weary wing,
Now deeming they are half-way there,
Perch on its boughs and sing.

THE TULIP TREE BLOSSOM.

And ere the sun full-robed appears,
He stops awhile to sup
The dew that gleams in his bright beams,
Within her golden cup.

I would that I, like thee, could live
Above the world as free,
As careless of its smiles or frowns,
FLOWER OF THE TULIP TREE.

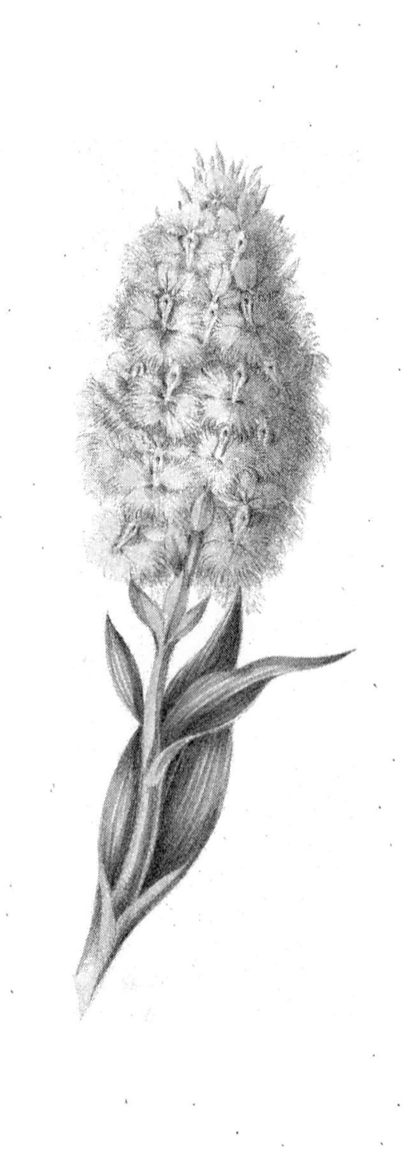

THE FRINGED ORCHIS.

This flower is lovely to the sight,
As any flower can be,
And yet 'tis not its beauty bright,
Which makes it dear to me.

It is because the one who gave
The treasure to my hand,
Hath given his frail form to the grave,
And sought the better land.

It is because he plucked it when
He was so near to heaven,
I almost fancied even then,
'Twas by an angel given.

TO THE WOOD LILY.

Lone Lily, on thy slender stem,
Thou'dst grace a regal diadem,
But art thou, lovely child of light,
As vain and proud as thou art bright?

The question is no sooner sent
From my rash lips, than I repent;
I know that pride is not thy dower,
Thou unassuming rustic flower.

Thou'dst smile with just as sweet a grace,
Into the little beggar's face,
As if she were a prince's child,
And all the world upon her smiled.

Thou, by the One who sends the showers,
And gentle dews upon the flowers,
Art with his charity imbued,
Blessing the evil and the good.

TO THE RHODODENDRON.

I CHARGE thee, flower, of beauty born,
  Lift not thy head too high,
For, like the lowliest of thy race,
  Thou, too, wert born to die.

The Power that lifts thee to the sun,
  And bends thee to the gale,
Doth watch, with equal care and love,
  The Lily of the vale.

The liberal hand that touched thy cheek
  With love's enchanting hue,
Gave the Lobelia's dazzling red,
  The Violet its blue.

Then not, like man, proud man, look down,
  And curl thy lip in scorn
On the companions of thy path,
  Because more lowly born.

Seest thou the Rose, so modestly
  Her lovely head recline?—
And yet the Queen of flowers is she,
  Whose blushes equal thine.

Yet, midst thy leaves of evergreen,
  So sweetly hast thou blown,
But for the fragrance of the Rose,
  Thou might'st dispute the crown.

THE HAREBELL.

The lovely flower that blossoms here,
Starts first a smile, and then a tear;
Its modest beauty charms the eye,
Its fragile form awakes the sigh.
To-day it swings its purple bells,
With every passing breeze that swells;
But seek it on another day,
Its fairy form has passed away.

So doth the infant's lovely smile,
From every thought of care beguile,
While to the mother's loving breast,
Its cherub head is warmly pressed.
No words to her fond tones reply,
He answers with his soft blue eye.
His ruby lip, the smile that curls,
Reveals two little shining pearls.

His tiny hands are stretched to clasp
The hand that takes them in its grasp.
What is there, in a sight like this,
To wake a thought of aught but bliss?
To-morrow, I will lead you where
That mother kneels in silent prayer,—
Her baby's soft blue eye is hid,
Beneath its moveless fringed lid.

THE HAREBELL.

Those ruby lips no more will part
In smiles to cheer her lonely heart;
No more those dimpled fingers press
Her breast, in all their loveliness.—
The flower lies safe, beneath the mould,
Protected from the Winter's cold;
The child's sweet spirit lives on high,
Where flowers and children never die.

HE SPARED MY FLOWER.

ONCE on a glowing Summer's day,
I saw a bunch of flowerets gay,
And opening buds, a rich display,
The tall grass waving round it.
And near it swept the mower's arm;
But such its native power to charm,
He could not do my floweret harm,
But left it where he found it.

And there for many a day it staid,
In tints of golden light arrayed,
And with the laughing sunbeams played,
Brightly and gaily dancing.
And every one who pass'd, it blessed,
While not a thought of care oppressed,
Or fear disturbed its gentle breast,
From scythe in sunlight glancing.

And thanks unto the gentle lad,
Who such an eye for beauty had,
In Nature's glowing colors clad,
With grateful heart I'd render.
But many suns have set, alas!
Since he from this fair world did pass,
And other men have mowed the grass,
But none with hearts more tender.

Yet could I find his lowly bed,
I'd plant a willow at its head,
Which o'er him should its branches spread
    Through all the Summer hours.
And as the twigs delighted swing,
And o'er him quivering shadows fling,
While all around the wild birds sing,
    I'd strew it thick with flowers.

THE WILD ROSE.

I LOVE the Rose of Summer,
  I love the sweet WILD ROSE,
I love the stream beside it,
  That softly rippling flows.
I love the month that brings it,
  The merry laughing June,
I love the bird that sings it,
  A gay and loving tune.

It has the sweetest fragrance
  Of any flower that blows;
The crown, there's no disputing,
  Belongeth to the Rose.
The garden rose is richest,
  For Flora's diadem,
But charming is the WILD ROSE
  For her a bosom-gem.

TO THE CARDINAL FLOWER.

How shall I paint the dazzling red,
That decks thy high resplendent head?
There's not a flower in field or lea,
In brilliance can compare with thee.
As well portray the diamond gleam
Of noonday on the rippling stream,—
As well attempt the rosy glow
That sunset gives to clouds of snow.
The Cardinal, his scarlet crown,
May proudly at thy feet lay down,
For glory's added to his power,
By giving name to such a flower.
I said I'd paint it, but whene'er I tried,
Its matchless hue, my feeble skill defied.
Go, hold aloft, said I, thy glowing light,
And still shine on, where nothing else is bright.
Go, bless the plants in humbler garb arrayed,
And give their low, damp homes thy cheering aid.
For, brilliant one, the happy power is thine,
To make the dismal swamp's dank margin shine.
Go, raise thy head to Him who lives in heaven,
And praise Him for the splendor he has given.—
A glimmering lamp at mournful midnight shone,
Shamed by the scene its faint beams rested on,—
A scene of deathless love, of deathlike life,—
A drunken father, and a dying wife.
In Charity's sweet form an angel creeps,
Through the low door, and o'er the sufferer weeps—

Holds to the dying lips the cooling draught,
And bids the noiseless fan soft breezes waft,—
Hushes with gentle voice the infant's moans,
With ready ear attends the sufferer's groans;
Bends low her form above the frameless bed,
And tells of Him whose blood for her was shed.—
To picture true, the spirit-beauty there,—
Her eyes on heaven, her parted lips in prayer,—
Has never been to mortal limner given,
But angels hung it on the walls of heaven.

AUTUMN DAYS.

O, SAY not Autumn's lovely days
Are sad, and full of gloom,
That all the tunes her wild harp plays,
Are "marches to the tomb."

Though, with her nimble fingers, May,
The green earth strews with flowers,
And June, in all her bright array
Of roses, decks her bowers,

Not Summer, with her richest gems,
And all her golden sheaves,
Can rival, on their bending stems,
Our frost-tipped Autumn-leaves.

Go seek the woods, and see the sun,
Through quivering leaves look down;
No need his fiery glance to shun,
He vails his burning frown.

For now o'er all the earth and sky,
A pensive beauty glows,—
A softened radiance from on high,
No other season knows;

While every quiet leaf that drops
So silent through the air,
And all the tree-crowned mountain tops,
A glowing beauty wear.

AUTUMN DAYS.

And if, with melancholy fraught,
  They touch the pensive heart,
'Tis only for the painful thought,
  They must so soon depart.

There is a bird, whose sweetest strains
  With her last breath are passed;
So Nature, of her hues, retains
  Her loveliest till the last.

The sweetest children soonest die;
  The saint's triumphant song,
Ne'er pours such heavenly melody,
  As from his dying tongue.

Then linger, lovely Autumn days,
  Our hearts with peace to fill,
To soothe us with thy softened rays,
  O, linger, linger still.

THE WILD ASTER.

When Autumn winds, with mournful tone,
  And many a fitful sigh,
Speak of the Summer pass'd and gone,
  And dry leaves round us lie,

We mourn to see the faded flowers,
  All withering on the stalk,
And we too sigh, when through their bowers,
  In pensive mood, we walk.

We grieve to think the snows will fall
  O'er each once lovely head,
And spread a cold white funeral pall
  Upon their lowly bed.

But while we sigh, a lovely gem
  Attracts our wondering sight,
And clustering round their parent stem,
  Is many a form of light.

And till November's cruel theft
  Each trembling leaf shall tear,
Which fair October's fingers left,
  They'll gleam in beauty there.

Yes, like an old and trusty friend,
  Who, when misfortune lowers,
Will still in kindness o'er us bend,
  And drop his tears with ours,

The lovely Aster wears a smile,
  That drives our cares away,
And says, the flowers but rest awhile,
  To bloom another day.

THE FRINGED GENTIAN.

I've watch'd, as eve led on the solemn night,
Star after star gleam forth with modest light,
Till, bolder grown, as more and more advance,
All heaven hath seem'd with one great joy to dance.
Soon, as by magic led, dark clouds approach,
With envious haste, and on the stars encroach;
With deepening gloom, they fast and faster come,
Join in one band, and cover all the dome.
As droops my head, in disappointment sore,
I heave a sigh for one more pleasure o'er.
The clouds in pity seem my grief to view,
Part their dark folds, and let one star shine through.
It blinks a moment, then creeps back again;
I close my curtain, and no more complain.

So Nature, viewing, as the Autumn led,
One after one, her favorite blossoms fled,
Saw the dead leaves o'er grove and valley lie,
Watch'd in her arms, the last pale Aster die,
Hung her wild harp upon a leafless tree,
And listened to its mournful melody.
The earth in pity heard the sad lament,
And to her arms this lovely blossom sent,
With fringes dipped in that imperial dye,
Which she had borrowed from the clear cold sky.
Nature beheld with joy the golden flashes
Smile brightly on her through their trembling lashes.
With grateful heart, the lone sweet flower she blest,
Then laid her down beneath the snows to rest.

A GRAVE AMONG THE FLOWERS.

If I should die when skies are clear,
In the sweet Spring-time of the year,—
O, hide me not, where lock and key
Shall shut the pleasant world from me;
But lay me where my narrow bed
May with fresh sprouting turf be spread,
From which the dew, and sun, and shower
Shall bring to life the budding flower.

And if in Summer's glowing heat,
This heart of mine shall cease to beat,—
O, lay me where the tall grass waves,
With bending flowers, o'er humble graves.
I'd rather have my lowly bed,
With late and early wild flowers spread,
Than covered with the costliest stone
That mourner's tears e'er dropp'd upon.

And if when Autumn mourns her bowers,
Despoiled of all her choicest flowers,
My earthly pilgrimage shall close,—
O, let my wasting form repose,
Where must the slow procession pass,
O'er dead leaves in the crisping grass,
Which whisper, as the flowers decay,
"So all that lives must pass away."

A GRAVE AMONG THE FLOWERS.

And should my soul be call'd to go,
When WINTER'S tempests fiercely blow,—
Still dig my grave, and let me lie,
Beneath God's open, azure sky.
Then, as the Spring's returning rain
Shall start the flowers to life again,
Some friend may smile through tears, and say,
"She'll bloom, when flowers for aye decay."

New York Botanical Garden Library
QK98 .B31
gen
Bauger, C. M./Wild flowers drawn and col
3 5185 00093 8264

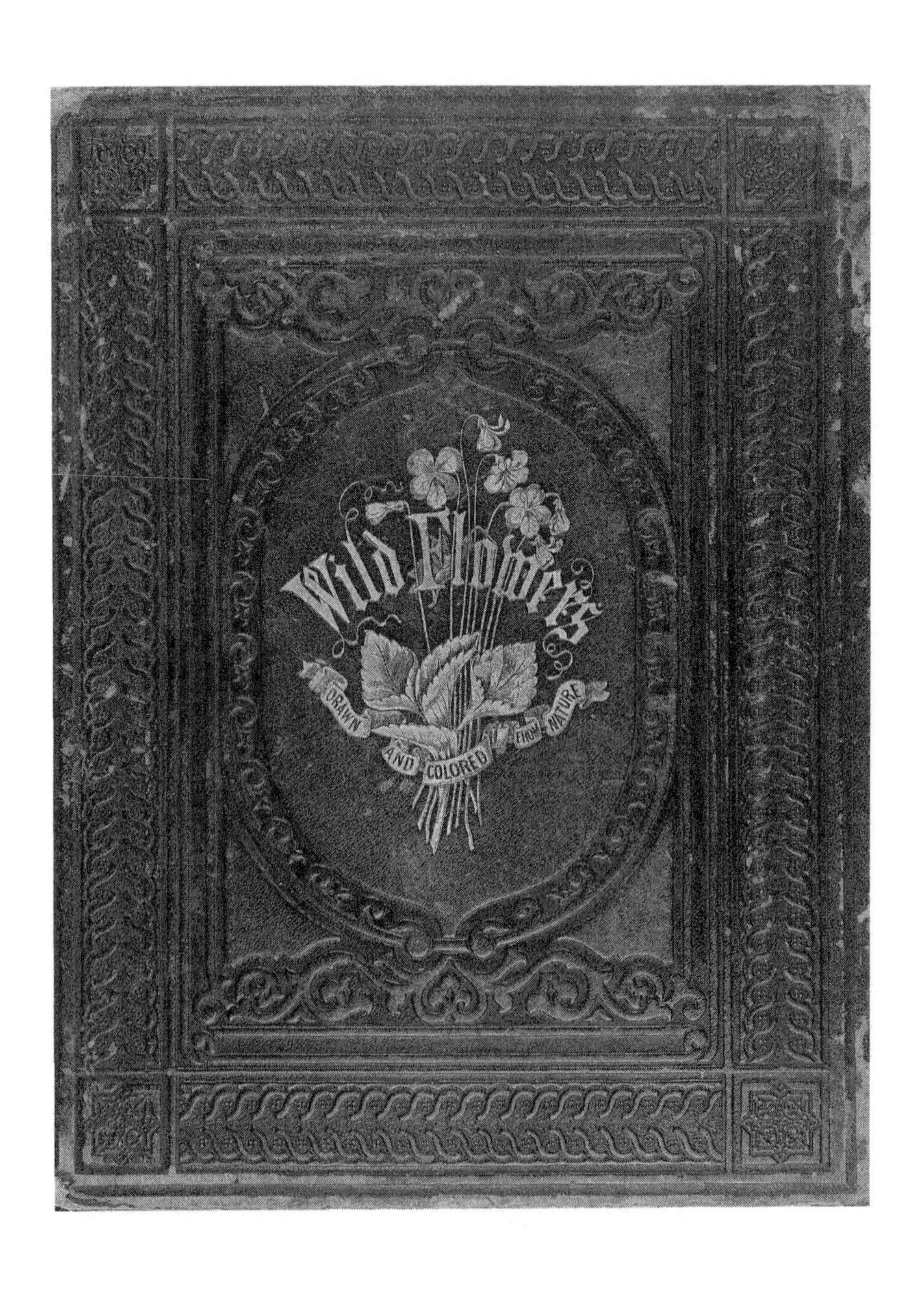
Wild Flowers
DRAWN AND COLORED FROM NATURE

CPSIA information can be obtained
at www.ICGtesting.com
Printed in the USA
BVHW052329170522
637236BV00004B/176